INTO Wild Tanzania

BLACKBIRCH PRESS

An imprint of Thomson Gale, a part of The Thomson Corporation

THOMSON GALE™

Detroit • New York • San Francisco • San Diego • New Haven, Conn. • Waterville, Maine • London • Munich

THOMSON
GALE

For more information, contact
Blackbirch Press
27500 Drake Rd.
Farmington Hills, MI 48331-3535
Or you can visit our Internet site at http://www.gale.com

Photo credits: cover, all pages © Discovery Communications, Inc. except for pages 6, 22, 25-26, 35, 38, 39 © Photos.com; pages 19, 42-43 © Corel Corporation

LIBRARY OF CONGRESS CATALOGING-IN-PUBLICATION DATA

Into wild Tanzania / John Woodward, book editor.
p. cm. — (The Jeff Corwin experience)
Includes bibliographical references and index.
ISBN 1-4103-0249-0 (alk. paper) — ISBN 1-4103-0250-4 (pbk. : alk. paper)
1. Tanzania—Description and travel—Juvenile literature. 2. Zoology—Tanzania—Juvenile literature. 3. Natural history—Tanzania—Juvenile literature. 4. Wilderness areas—Tanzania—Juvenile literature. 5. Corwin, Jeff—Travel—Tanzania—Juvenile literature. I. Woodward, John. II. Corwin, Jeff. III. Series.

DT440.5.I58 2004
590'.9678—dc22 2004004486

Printed in the United States
10 9 8 7 6 5 4 3 2 1

Ever since I was a kid, I dreamed about traveling around the world, visiting exotic places, and seeing all kinds of incredible animals. And now, guess what? That's exactly what I get to do!

Yes, I am incredibly lucky. But, you don't have to have your own television show on Animal Planet to go off and explore the natural world around you. I mean, I travel to Madagascar and the Amazon and all kinds of really cool places—but I don't need to go that far to see amazing wildlife up close. In fact, I can find thousands of incredible critters right here, in my own backyard—or in my neighbor's yard (he does get kind of upset when he finds me crawling around in the bushes, though). The point is, no matter where you are, there's fantastic stuff to see in nature. All you have to do is look.

I love snakes, for example. Now, I've come face to face with the world's most venomous vipers—some of the biggest, some of the strongest, and some of the rarest. But I've also found an amazing variety of snakes just traveling around my home state of Massachusetts. And I've taken trips to preserves, and state parks, and national parks—and in each place I've enjoyed unique and exciting plants and animals. So, if I can do it, you can do it, too (except for the hunting venomous snakes part!). So, plan a nature hike with some friends. Organize some projects with your science teacher at school. Ask mom and dad to put a state or a national park on the list of things to do on your next family vacation. Build a bird house. Whatever. But get out there.

As you read through these pages and look at the photos, you'll probably see how jazzed I get when I come face to face with beautiful animals. That's good. I want you to feel that excitement. And I want you to remember that—even if you don't have your own TV show—you can still experience the awesome beauty of nature almost anywhere you go—any day of the week. I only hope that I can help bring that awesome power and beauty a little closer to you. Enjoy!

Best Wishes!
Jeff

INTO Wild Tanzania

From an ancient crater that teems with life to the legendary Serengeti, come with me and watch Africa come alive.

I'm Jeff Corwin.
Welcome to Tanzania.

Welcome to Tanzania...where zebras roam the Serengeti and Africa really comes to life.

I've just been awakened by a snake! A nice discovery indeed—it's a cobra. I imagine for most people, being awakened by a snake in the middle of the African bush would be a nightmare. But for me, this is heaven on earth. This is an Egyptian cobra. It's showing its very dramatic hood, which it spreads to appear larger than life.

As with all cobras, it belongs to the Elapidae family. What most elapids share in common is that they produce a neurotoxic venom in glands behind their jaws. That venom is then funneled down tubes that are connected to fixed

fangs. It likes to eat amphibians. You can find this creature along waterways from Egypt to South Africa. I'm going to release this snake so we can start our journey.

This part of Tanzania is a wildlife melting pot, second to none in the world. There's a different creature and danger lurking around every corner. I'll show you as much of this amazing place as I can. Tanzania is a gold mine for unique and sometimes dangerous snakes like these. So I guess you could say we've just struck the mother lode.

You can tell that there was water running through this river wash, perhaps a week ago. But the sun's been drying it out, baking it into a cake. There's something moving through this grass. It could be a monitor or a turtle. Check this out! Had he not tickled this grass a wee bit, we never would have known that hidden in this tuft of greenery is a colossal snake. In fact, it's the largest species of snake in this part of the world.

The more I do this stuff, the more I stress out. What freaks me out is that it could be anywhere. And a snake like this is large enough to swallow a small antelope. As it gets larger throughout its life, it could eat even larger creatures.

This animal is not venomous, though. I've been bitten by pythons before, and when they're this big, it really, really hurts. He has about a four-foot strike range. I'm going to try to do an old trick. I'm going to try to muffle his head with my shirt.

I've got the python in my hands. It feels like he's breaking my thumb! This is an African rock python with big teeth. The way I captured it, I put as little stress on the animal as possible. I used the old matador technique by distracting him with the shirt and waiting for an opportunity for me to sneak my hand in and secure him. What I'm trying to avoid is those sharp teeth. He'll bite you if you get too close. It's not venomous, but you're talking about a quarter-inch-long, razor-sharp tooth.

After these snakes secure the prey with their teeth, constriction begins. Lucky for me, he's only constricting me slightly, and not around my neck. This snake isn't even full size, but it's about ten feet long and weighs forty pounds. How big do they get? This is a snake that can reach twenty to twenty-five feet in length—so big that it can swallow an antelope. There are even stories of them swallowing young cape buffalos. Now would he try to eat me?

Unlikely. But imagine if I were a small antelope. I would have walked right into my death as this hidden animal waited for food. It's an ambush predator. What if it was one of those twenty footers? I would have been an appetizer!

Uh oh, he's trying to coil back to strike, so I'm going to let him go. He's tired of me. And you know what? I'm a little tired of him too, so let's place him back. There you go, African rock python.

Unfortunately, poaching is still a very real problem in this part of the world. The richness of the wildlife attracts those who kill for profit. A Maasai ranger named Thomas Kuya has offered to give me a firsthand look at their handiwork. He's going to show us some wild animals. Thomas is armed with a rifle because his job is to keep poachers off this land. All of the animals that live here are at risk. The poachers are deadly serious about taking animals, so Thomas is deadly serious about stopping them. He's looking for metal wires, snares. He just found one.

The poacher used some tree branches to hide the snare. When the animal walks by, his foot is caught and he's tied to the tree. Thomas has captured poachers in the act of poaching. Poachers can get up to seven years or more in jail. If you're in a national park in some parts of Africa, you risk your life poaching. Now there's one less snare to cause damage in this very special ecosystem.

Check this out! This isn't super fresh, but it sure is a good sign that we are in elephant territory. You've heard of a road apple. Well, this is a giant mammoth apple, literally, from the living descendants of a mammoth, elephant dung.

There has been a lot of activity here. You can see where these giants have come in and pushed down the vegetation in search of

food. When we think of elephants, we usually think of gentle giants. But elephants can be dangerous. You never want to mess with a bull that is in musk. And mothers will stop at nothing to defend their calves. Even though we're only here to see these creatures and learn from them, if we get too close, we'll be squashed. We have to respect these animals.

Way off in the distance, there is a brown mound. It's not a termite mound. That's a mound of pachyderm, of elephant. Let's see how close we can get. Wow! That's one big elephant. He could squash you easily. He probably weighs twelve thousand pounds.

We've now been tracking these animals for quite a while. We've gone over some serious terrain, and we're finally close. Look at that bull. He's huge! You're probably wondering why this creature is alone. Is it a lonely bull elephant? That's not the case at all. When a male elephant turns from nine to fourteen years old, he's kicked out.

This guy is huge! He probably weighs twelve thousand pounds.

The head matriarch will push him out. He might join up with a bachelor herd of adolescent bulls. But when he gets this big, when he weighs ten thousand pounds, he likes life on his own.

Thomas has just spotted a mother. Her baby stands six feet high. It weighs two thousand pounds and it's only two or three years old. Uh, oh. Don't move. Mom sees us. Hopefully, she's not going to charge. If she charges, you don't want to run, because she'll charge right after you. We're going to start moving. Okay. There, she's walking off. I've got a little bit of the shakes here. It was an adrenaline rush. There was adrenaline running in my blood and the blood of these great animals. These elephants are extraordinary mothers. They will put their own lives or your life on the line to defend their babies. Elephants are complex, intelligent, powerful creatures. They are worthy of admiration, respect, and a lot of space.

Now I'm with my friend Sadik at Ngorongoro Crater. It's only twelve miles wide, and it is literally bubbling over with wild animals. We're in luck, because during migration, when the wildebeests and zebras move through here, as they are doing now, predators show up.

We've got a mother cheetah and two baby cheetahs. Let's slowly creep in and see how close we can get. They're hunting. Look at this—she's going for it. The cheetah's going to get that gazelle. I've seen this twice in my life, and each time, I have mixed emotions. Part of me is excited, the hunter part of me. That part gets riled up. The other part of me feels a little sad, because a beautiful, healthy animal has fallen prey, and its life is over. But that's the way it works.

Now she's got the kill by the head and the babies want it. The babies are hungry, and now she's going to feed them. They have to eat quickly. There is no time to waste, because waiting in the wings all around them are other predators that will slowly move in to take that kill away. Cheetahs are built for speed. They're built to fly like a bullet, take down the prey, and eat it quickly. They are not built to hang around and defend. A cheetah that chooses to hang around and defend its kill from something like a leopard or a lion will be a dead cheetah.

Young cheetahs face many challenges growing up. They have to deal with competition, predators, and lack of food and water. All these things make their survival rate very low. Life is tough for a cheetah. Less than half of all cheetahs survive

Young cheetahs have to overcome many challenges to reach adulthood.

to adulthood. And even then, their life expectancy out in this wilderness is at most seven years. But today is a good day for these cheetahs. Their bellies are full, there's no competition, and the sun is out. It was a bad day for the gazelle, but a good day for the cheetahs.

Today is our lucky day, because we're about to meet one of the rarest mammals living on our planet. We're going to go in slowly, because if we go too fast, we'll startle these animals. And if they're startled, they will charge.

I can't believe we are so close to these animals! In front of us is a large example of what is left of this species, the black rhino. Rhinos are extremely territorial, especially the males. They let other rhinos know where their territory is by establishing a dung heap. What they'll do is pile up their scat and micturate, or urinate, in the region as well. There can be dung heaps that are six or seven feet high, giant mounds that basically say, "You're in my property."

This creature's head terminates in a horn that almost curls up above the forehead. And then underneath that, they've got a second horn to the back. The bigger the horn, the more

impressive they are to the females.

Now his tail is up in the air and he's charging forward. That's the sign of a black rhino that is either ticked off or nervous. These animals can move up to thirty-five miles an hour—not for a long period of time, but plenty of time to get from there to here in a few seconds, and for us to be upside down. These are amazing beasts. Tragically, they are close to extinction, but it's not over yet. There's still a chance that these animals could be in our future. It's just going to take a lot of work.

Just three miles from those rhinos, the king of beasts is waiting. Look at this beautiful lion. We're not at a zoo—this is the real deal. There is a great wind howling over the grasses. The environment is humming with wind, excitement, and energy. The African lion is the largest terrestrial predator in Africa. Females weigh around 300 to 350 pounds. Males weigh 400 to 500 pounds.

The African lion is truly king here.

The lion is sitting on his throne, a natural bluff. This gives him the advantage of being able to see potential prey through the valley. He's keeping an eye out on those wildebeests, looking for one that's a little lame. He's also keeping an eye out for his buddies. This is what being in a pride is all about. Most likely, these two are brothers.

To understand how these animals are able to position themselves with this great piece of real estate, you've got to know a little bit about lion society. The leaders of this pride are these lions. Below them are the females. There can be from two to twenty lionesses. When life is good, the lion has his real estate. He's got his lionesses out there catching his food. He and his brother or him by himself ward off the other lions moving through this habitat. But if he becomes a little weak, a battle occurs.

Two brothers might come in and drive off the king from his throne. Perhaps he's wounded. Now he sits out there on the outskirts, lower than prey, watching as these invaders pillage all he has built. They not only take over his real estate, but they kill his cubs. It brings the females into estrus. They will breed with the invaders. This is survival of the fittest. It's raw and it's ugly, but it's real and it's Africa. Now we're going to leave Ngorongoro and head out across the Serengeti.

This is the village of Mto Wa Mbu. Each week people gather to share gossip and tell stories. It's also a place of commerce. And the people that are doing this are the Maasai. The Maasai people are closely tied to the land. They come together in this market to share stories and sell their goods.

I just love the smells, the activity. This place is a celebration of the Maasai community. Here, you can trade for traditional medicines harvested from the bush. For these people, the environment is their pharmacy.

This is the material Maasai use to paint their faces. It's made out of soil. I'm going to buy some. With my new purchase in hand, it's time to continue my exploration of Tanzania.

Now we're at the Serengeti. The Maasai call it *Serenget*, or the place where the land moves on forever. These twelve thousand square miles have remained virtually unchanged for millions of years.

This is a balanite tree that's been charred by a fire that worked through this field perhaps ten years ago. What's exciting about this tree is what's at the top. That branch is moving. That's because it's a snake! This is a boomslang. It's a snake you have to be very careful with because it is very venomous.

This creature has served African myth and legend for hundreds of years. What's interesting about this snake is that the fangs are to the back of the mouth. Many people think because it's rear-fanged, you have a less chance of being injured by this animal. But that is not true. Because he opens his mouth at a ninety-degree angle, and his back fangs are extremely long. He reaches out, grasps, and then injects a powerful toxin.

When you're bitten by this snake, it's not as painful as many other snakebites. You might think, "I got bitten, but I'm feeling okay." And then at lunch and you're laughing about the story. You feel great. And then, twenty-four hours later, you start to develop symptoms. Your gums start to dissolve, and you start to weep blood on the inside and outside. Essentially, you melt. That's how the venom works. You need to get antivenin quickly.

Isn't this great? We're surrounded by giraffes. A giraffe looks like its moves very slowly. But in fact, the word *giraffe* comes from an Arabic word, *zirafah*, that means "one who moves swiftly." This creature moves swiftly on those long legs. From hoof to shoulder, this creature

measures ten feet. Overall, it's eighteen feet tall. It's deceiving when it moves, because those long legs seem to be moving in slow motion, but in fact, they're moving very fast.

The giraffe is an amazing creature. It has an eighteen-inch-long prehensile tongue. It reaches out and wraps around the vegetation, twisting just like a finger, pulling up the most succulent part of the leaf.

Look at the top of this animal's head. It has fuzzy horns. But despite their fuzzy, soft appearance, they're an excellent form of defense. They're also used in mating ritual between combative males trying to get access to females. In fact, it's called necking. They'll sway their heads back and forth banging into each other. These animals have been killed by that behavior. Skulls have been smashed, vertebra have been shattered when one neck hits the other.

Giraffes may look a bit odd, but they're perfectly designed.

Look at those two-inch-long eyelashes! They're excellent defense against ants when these creatures are grazing on acacia trees. They also defend against thorns. When that eyelash rubs against a thorn, the giraffe detects it and move its head so it can eat the vegetation without risking an eye injury. At first glimpse, the giraffe appears a bit odd. But when you look at it up close, you understand this creature is very well designed.

If you really want to get a good feel for a region, you've got to experience it the way the locals do—the food and the lodging. We're also traveling the way the locals do. For just a few cents, anyone can hitch a ride on one of these *mutatus*. It's cheaper than the subway in Boston. And it's not as crowded.

We've got some activity going on here. I think it's a dead wildebeest. Here's a white-capped vulture. Over there is a lappet-faced vulture. They come together along with the storks and the hyenas

Vultures aren't very pretty, are they?

and the jackals to serve as the sanitary keepers of the Serengeti. It's kind of an ugly buffet to witness. But keep in mind, this is just one example of how energy flows through this great and exciting ecosystem.

Look how this carcass has been cleaned from the inside out. All the flesh has been eaten. The ribs have been picked of every bit of flesh. The inside is all cleaned out. That's why these vultures don't have any feathers on their heads, so they won't get dirty.

This is an excellent example of a Maasai village. I'm looking for a Land Rover or a bus or something. I probably should have stayed with the Land Rover or

Passing through a Maasai village

I couldn't pass up this challenge.

Riding a bicycle across the Serengeti. It doesn't get much better than this!

even the *mutatu*, but I couldn't resist another challenge. How many times do you get a chance to ride a bicycle across the Serengeti?

You can see Mount Kilimanjaro behind the acacia trees.

Look at that view! You know, back home in New England, during those cold winter days when I want to escape, I visualize this. I think of memories of past experiences in Africa. This sweeping savanna of grass is being kissed by a warm wind, and it's dotted by these old, twisted acacia trees. And off in the distance, there is the mighty beast of rock, Mount Kilimanjaro. And somewhere in those foothills is the place we've come to explore, Ndarakwai, a place of adventure, a place of wildlife.

Getting refreshed at safari camp

Finally, we've reached our safari camp. This is Ndarakwai. Nighttime out here in the savanna can be a bit tricky, especially if you don't have the energy to keep your wits about you. So I'm going to bed. When I wake up, I've got ten thousand acres of wonderful habitat waiting for exploration. So, see you in the morning.

This adorable baby is an orphan.

Now this isn't just any ordinary African camp. The people here at Ndarakwai specialize in wildlife rehabilitation, and they've adopted an orphaned elephant. She was found alone and starving. Her mother was most

likely a victim of ivory poachers.

Now I know what this bottle's for. She might be a baby, but I certainly would never want to take her bottle away. This was four pints of fluid, and she sucked it down in two seconds! And she's going to get every last drop out of that bottle. She is probably around two years old. When this creature came into the world after twenty-two months of pregnancy, it weighed about a couple of hundred pounds.

From a year onward, this creature starts to transition out of mother's milk to solid food. But nursing isn't just cut off immediately. In fact, these animals can be nursing up to four years. I read of one calf that nursed for ten years. In any case, the responsibility of taking care of a baby elephant is a big one indeed, because it's not very easy.

Well, I'm afraid our voyage is at an end. But I can't think of a better way to end our safari in Tanzania than by following an elephant to our next wildlife adventure.

Glossary

amphibian a class of cold-blooded organisms including frogs and toads

ecosystem a community of organisms

elapid a type of venomous snake

micturate urinate

poaching illegally hunting an animal, especially an endangered one

prehensile adapted for grasping

prey the animal hunted by a predator

pride a group of lions

safari an overland expedition, especially one for hunting or exploring in Africa

savanna a tropical or subtropical grassland

terrestrial living on the land

toxin poison

venom a poison used by snakes to attack their prey or defend themselves

venomous having a gland that produces poison for self-defense or hunting

Index